SIXTH GRADE DAILY GEOGRAPHY: SIMPLE GEOGRAPHY LESSONS

WONDERS OF THE WORLD FOR KIDS
6TH GRADE BOOKS | CHILDREN'S MYSTERY & WONDER BOOKS

BABY PROFESSOR
EDUCATION KIDS

Speedy Publishing LLC
40 E. Main St. #1156
Newark, DE 19711
www.speedypublishing.com

Copyright 2015

All Rights reserved. No part of this book may be reproduced or used in any way or form or by any means whether electronic or mechanical, this means that you cannot record or photocopy any material ideas or tips that are provided in this book

Geography is a field of science dedicated to the study of the lands, the features, the inhabitants, and the phenomena of Earth.

The Amazon rainforest produces more than 20% the world's oxygen supply.

Ninety percent of the world's ice covers Antarctica.

The Vatican city is the smallest internationally recognized independent state in the world.

Glaciers store between 70% and 80% of all the freshwater on the planet.

The Himalayas is home to nine of the ten highest peaks on Earth.

Budapest was originally two separate cities, Buda and Pest. The 2 areas are separated by the Danube River.

Iceland is home to around 200 volcanoes and it has a third of all lava flows found on Earth.

Canada has more lakes than the rest of the world combined.

Siberia contains more than 25% of the world's forests.

The Nile river is generally regarded as the longest river in the world.

Monaco is the most densely populated country in the world.

Angel Falls is the world's highest uninterrupted waterfall, with a height of 979 meters.

The Namib desert, at 80 million years, is the world's oldest desert.

Mount Everest is the highest mountain in the world, At 8848 meters.

Made in the USA
Monee, IL
25 January 2023